THIS BOOK BELONGS TO:

Careers in Tech: Certification Exam Guide

Weekly Goals

TO DO:

- [ ]
- [ ]
- [ ]
- [ ]
- [ ]
- [ ]
- [ ]
- [ ]
- [ ]
- [ ]
- [ ]
- [ ]
- [ ]
- [ ]
- [ ]
- [ ]

COMPLETED:

NOTES

Weekend Program

WORK PLAN:

RATE & IMPROVEMENT:

NOTES

MONDAY:

TUESDAY:

WEDNESDAY:

THURSDAY:

FRIDAY:

SATURDAY:

SUNDAY:

NOTES

NOTES

You Can Do it!

NOTES

NOTES

NOTES

NOTES

NOTES

NOTES

NOTES

NOTES

NOTES

NOTES

Weekly Goals

TO DO:

- [ ]
- [ ]
- [ ]
- [ ]
- [ ]
- [ ]
- [ ]
- [ ]
- [ ]
- [ ]
- [ ]
- [ ]
- [ ]
- [ ]
- [ ]
- [ ]

COMPLETED:

NOTES

Weekend Program

WORK PLAN:

RATE & IMPROVEMENT:

NOTES

MONDAY:

TUESDAY:

WEDNESDAY:

THURSDAY:

FRIDAY:

SATURDAY:

SUNDAY:

NOTES

NOTES

You Can Do it!

NOTES

NOTES

NOTES

NOTES

NOTES

NOTES

NOTES

NOTES

NOTES

NOTES

Weekly Goals

TO DO:

COMPLETED:

NOTES

Weekend Program

WORK PLAN:

RATE & IMPROVEMENT:

NOTES

MONDAY:

TUESDAY:

WEDNESDAY:

THURSDAY:

FRIDAY:

SATURDAY:

SUNDAY:

NOTES

NOTES

You Can Do it!

NOTES

NOTES

NOTES

NOTES

NOTES

NOTES

NOTES

NOTES

NOTES

NOTES

Weekly Goals

TO DO:

COMPLETED:

NOTES

Weekend Program

WORK PLAN:

RATE & IMPROVEMENT:

NOTES

MONDAY:

TUESDAY:

WEDNESDAY:

THURSDAY:

FRIDAY:

SATURDAY:

SUNDAY:

NOTES

NOTES

You Can Do it!

NOTES

NOTES

NOTES

NOTES

NOTES

NOTES

NOTES

NOTES

NOTES

NOTES

Weekly Goals

TO DO:

- [ ]
- [ ]
- [ ]
- [ ]
- [ ]
- [ ]
- [ ]
- [ ]
- [ ]
- [ ]
- [ ]
- [ ]
- [ ]
- [ ]
- [ ]
- [ ]

COMPLETED:

NOTES

Weekend Program

WORK PLAN:

RATE & IMPROVEMENT:

NOTES

MONDAY:

TUESDAY:

WEDNESDAY:

THURSDAY:

FRIDAY:

SATURDAY:

SUNDAY:

NOTES

NOTES

You Can Do it!

NOTES

NOTES

NOTES

NOTES

NOTES

NOTES

NOTES

NOTES

NOTES

NOTES

Weekly Goals

TO DO:

COMPLETED:

NOTES

Weekend Program

WORK PLAN:

RATE & IMPROVEMENT:

NOTES

MONDAY:

TUESDAY:

WEDNESDAY:

THURSDAY:

FRIDAY:

SATURDAY:

SUNDAY:

NOTES

NOTES

You Can Do it!

NOTES

NOTES

NOTES

NOTES

NOTES

NOTES

NOTES

NOTES

NOTES

NOTES

Weekly Goals

TO DO:

COMPLETED:

NOTES

Weekend Program

WORK PLAN:

RATE & IMPROVEMENT:

NOTES

MONDAY:

TUESDAY:

WEDNESDAY:

THURSDAY:

FRIDAY:

SATURDAY:

SUNDAY:

NOTES

N O T E S

You Can Do it!

NOTES

NOTES

NOTES

NOTES

NOTES

NOTES

NOTES

NOTES

NOTES

NOTES

Weekly Goals

TO DO:

COMPLETED:

NOTES

Weekend Program

W O R K   P L A N :

R A T E   &   I M P R O V E M E N T :

NOTES

MONDAY:

TUESDAY:

WEDNESDAY:

THURSDAY:

FRIDAY:

SATURDAY:

SUNDAY:

NOTES

NOTES

You Can Do it!

NOTES

NOTES

NOTES

NOTES

NOTES

NOTES

NOTES

NOTES

NOTES

NOTES

Weekly Goals

TO DO:

- [ ]
- [ ]
- [ ]
- [ ]
- [ ]
- [ ]
- [ ]
- [ ]
- [ ]
- [ ]
- [ ]
- [ ]
- [ ]
- [ ]
- [ ]
- [ ]

COMPLETED:

NOTES

Weekend Program

WORK PLAN:

RATE & IMPROVEMENT:

NOTES

MONDAY:

TUESDAY:

WEDNESDAY:

THURSDAY:

FRIDAY:

SATURDAY:

SUNDAY:

NOTES

NOTES

You Can Do it!

NOTES

NOTES

NOTES

NOTES

NOTES

NOTES

NOTES

NOTES

NOTES

NOTES

Weekly Goals

TO DO:

COMPLETED:

NOTES

Weekend Program

WORK PLAN:

RATE & IMPROVEMENT:

NOTES

MONDAY:

TUESDAY:

WEDNESDAY:

THURSDAY:

FRIDAY:

SATURDAY:

SUNDAY:

NOTES

NOTES

You Can Do it!

NOTES

NOTES

NOTES

NOTES

NOTES

NOTES

NOTES

NOTES

NOTES

NOTES

Weekly Goals

TO DO:

COMPLETED:

NOTES

Weekend Program

WORK PLAN:

RATE & IMPROVEMENT:

NOTES

MONDAY:

TUESDAY:

WEDNESDAY:

THURSDAY:

FRIDAY:

SATURDAY:

SUNDAY:

NOTES

NOTES

You Can Do it!

NOTES

NOTES

NOTES

NOTES

NOTES

NOTES

NOTES

NOTES

NOTES

NOTES

Weekly Goals

TO DO:

- [ ]
- [ ]
- [ ]
- [ ]
- [ ]
- [ ]
- [ ]
- [ ]
- [ ]
- [ ]
- [ ]
- [ ]
- [ ]
- [ ]
- [ ]
- [ ]

COMPLETED:

NOTES

Weekend Program

WORK PLAN:

RATE & IMPROVEMENT:

NOTES

MONDAY:

TUESDAY:

WEDNESDAY:

THURSDAY:

FRIDAY:

SATURDAY:

SUNDAY:

NOTES

N O T E S

You Can Do it!

NOTES

NOTES

NOTES

NOTES

NOTES

NOTES

NOTES

NOTES

NOTES

NOTES

Weekly Goals

TO DO:

COMPLETED:

NOTES

Weekend Program

WORK PLAN:

RATE & IMPROVEMENT:

NOTES

MONDAY:

TUESDAY:

WEDNESDAY:

THURSDAY:

FRIDAY:

SATURDAY:

SUNDAY:

NOTES

NOTES

You Can Do it!

NOTES

NOTES

NOTES

NOTES

NOTES

NOTES

NOTES

NOTES

NOTES

NOTES

Weekly Goals

TO DO:

COMPLETED:

NOTES

Weekend Program

WORK PLAN:

RATE & IMPROVEMENT:

NOTES

MONDAY:

TUESDAY:

WEDNESDAY:

THURSDAY:

FRIDAY:

SATURDAY:

SUNDAY:

NOTES

NOTES

You Can Do it!

NOTES

NOTES

NOTES

NOTES

NOTES

NOTES

NOTES

NOTES

NOTES

NOTES

Made in the USA
Coppell, TX
04 November 2026

11282147R00109